No Big Bang

Publisher's Cataloging-in-Publication

Prideaux, David, 1944- author.
    No big bang : a three part look at a bent universe /
by David Prideaux.
    pages cm
    LCCN 2015911975
    ISBN 978-1-62901-282-7 (pbk.)
    ISBN 978-1-62901-283-4 (Kindle)

    1. Cosmology--Philosophy.   I. Title.

BD511.P75 2015   113
                              QBI15-600188

*Scan QR code to learn
more about this title*

Publisher: Inkwater Press | www.inkwaterpress.com

Paperback
ISBN-13 978-1-62901-282-7 | ISBN-10 1-62901-282-3

Kindle
ISBN-13 978-1-62901-283-4 | ISBN-10 1-62901-282-3

Printed in the U.S.A.

1 3 5 7 9 10 8 6 4 2

David Prideaux

no big bang

a three-part look at a *bent* universe

INKWATER PRESS

Portland•Oregon
inkwaterpress.com

Contents

Acknowledgments

My first words of thanks are to *all* of you — my fellow human beings and citizens, present and past. In the same way that it takes a village to raise a child, it also takes all of us to enable any one of us to succeed. So my list of names will have to be completely inadequate. Nevertheless, I feel I should try. If I have neglected to name a particular person who was vital to this book, I apologize.

My dear wife Alice has been supportive of me for nearly 40 years. My mom and dad "had my back" as I grew to know science as a youngster. And, I have a deep and reverent attitude toward *all* who tell truth, perhaps even to such an extent as to be willing to bravely sacrifice one's life itself in pursuit of bare naked truth and honest communication. I grew up Quaker, and the Friendly way guided me. Particular Friends such as a remarkable mentor, Howard Scott, showed me, by example, how to stand on principle.

As I have written this book, I have very often felt a recurring affinity with Charles Darwin. To try to invent something new "out of whole cloth"—especially when faced with opposition—is not an easy task. Good ideas don't grow on trees. And making sense may obligate one to endure pain, have persistence, and exercise patience. In a more general sense, *anyone* who struggles to overcome adversity was and is a direct inspiration for me.

Lastly, thanks to my helpful publisher, Inkwater. All images are from *www.pixabay.com*.

Introduction to All Three Parts

A quiet rebel looks into space.

Who is this gazer?

It is I. That's how it all began. I was a young boy peering at the moon through my own telescope. Dazzling magnification practically knocked me off my feet. A blazing moon appeared very close. The year was about 1956.

What in the world is going on out there?

I wanted to know. I wanted to come up with my own answers.

Out of that desire — my ambition to know — was born this book. It's an answer to a questing soul.

Why would someone question the Big Bang?

It might seem odd for someone to speak out against such a widely accepted theory as the Big Bang. Yet, many people like to be open-minded. Many are willing to consider alternatives to the status quo. I am counting on those folks. The person who pooh-poohs any critique of the Big Bang probably will not read further than this. I'm aware of that. It's OK.

Who am I to question the Big Bang?

Now — those who know anything about astronomy (a) will recognize in me a kindred spirit, but also (b) could have a mild objection (or could bristle with fierce indignation) if I'd dare assume authority to write about science without first having proved some academic credentials.

I do have some academic background (a master's degree in a different field), some credibility and some life research under my belt. But in the final analysis, I have a one-word answer. I would say to most types of questions about qualifications and validity: it's a matter of being *inherent* (intrinsic).

What I mean is: Does my idea ***inhere***? See if it can stand on its own feet. Test it. Examine it. Does my idea contain within itself believability, credibility or sensibility, ***just by its very nature***? Can you see any inherent authority if you take my idea out of context and consider it alone? What does it look like?

It follows from this argument that you'd have to read about it to get the gist of it. I respectfully and humbly ask you — the doubting, skeptical reader — to give it a go. Isn't that something science depends upon?

And if you need another reason to reconsider the Big Bang, think about how many kooky, goofy and perplexing notions are already out there. What's the harm of looking at yet one more idea?

If you'd like a terse summary, please look in the Epilogue at the conclusion of Part 1 of my book.

My aim is to simply get an idea circulating. I've not had any intention of "fleshing out" the idea with developing math and calculations that others would be much more qualified to do than I.

The value of seeing and looking anew

One way to answer my first question (What is going on?) is that it depends on how you look at it.

I now ask you to ponder something. It's a bit of a riddle. Robertson Davies said:

> The eye sees only what the mind is
> prepared to comprehend. (Davies)

In other words, we are pre-programmed. We are set up to make sense of things. We pre-select our perceptions. Training, experiences, beliefs, abilities and pre-existing mind frames choose for us what to notice. We see what we want to see.

The reverse is true too. We cannot and we will not see what is not within our expected fields of vision. Magicians know how to take advantage of this by staging tricks. They set us up to see illusions. By using sleight-of-hand, good communication skills, leading suggestions, and legerdemain, they can fool us.

Witnesses to a crime or a traffic accident may remember events differently. Dogs hear ultra-sound. Carnivores concentrate intensely and zero-in on prey animals. Bees interpret other bees' dancing motions. Perception is particular. Seeing is specific.

In science, one has to guard against chasing after false leads or following illusions. But sometimes we get stuck. We may find we're trapped in a cul-de-sac of thought. We may need to rebel against the present in order to create a future insight.

I am now making a direct appeal to the reader.

I want the idea of "no big bang" to be included in thinking about how we came to be, where we are going, and what we are. I will explain.

But first, I must acknowledge this is a radical departure from the accepted opinions of most of today's educated pundits, learned commentators, and experts.

Note that the general public doesn't necessarily look at it that way.

Keeping an Open Mind

So beware! *Caveat emptor!* One who goes against the opinions of the majority and the establishment risks the ire of those most challenged. One should expect resistance. Change often comes slowly.

Yet to take a chance (of offending our conventional wisdoms) is the only way an open society and modern science can progress, over the long run.

If a science is worth its salt it must listen to voices of dissent. A scientist must keep open to varying perspectives. Science must actively critique its own culture. Science asks questions. It poses puzzles. Especially—that intellectual responsibility should include "sacred cows."

The history of science shows us time and again that when we ask important, new, off-beat or odd questions, we open up the possibility of new discoveries. If the novel question is not asked, not heeded or not considered, then one's eyes will be unable to see beyond present and past. Those eyes, unseeing, will stay stuck in their owners' heads. Those heads will be OK with status-quo understanding.

My intended reader is firstly an expert scientist (perhaps an astronomer) who is willing to have an open mind.

But secondly, I hope the average lay reader can see in this book some provocative thoughts. I absolutely think *any* interested person's curiosity and brainpower can penetrate this subject.

As one good science writer put it in his book, *The Runaway Universe,*

> ...we must remain aware that new data and new interpretations may soon appear, causing us once again to question the framework within which we conceive the cosmos. (Goldsmith, p. 6)

The scientist who is also professional should humbly admit to the evolution of the "latest" truth.

The Role of Competition

In an ideal world, competition serves to make all parties and participants and ideas stronger. Strive to know. Use it or lose it (says a proverb from the past). If we put forth effort and compete, we will presumably develop ever better and more effective ways to succeed. In science and technology, this back-and-forth vying can get intensely personal.

Goldsmith said (once again in *The Runaway Universe*)

> ...competition to achieve new insights plays a key role in advancing knowledge...scientists...know that if they can disprove another's claims, their reputations will rise. This fact embodies the organized skepticism institutionalized in the world of science, the notion that only by surviving harsh criticism from experts can a new result gain respect. (Goldsmith, pp. 113–114)

Other Critiques of the Big Bang

Eric J. Lerner, Saurya Das and others have suggested evidence not supportive of the Big Bang. Mine is certainly not the only "no big bang" argument out there. The unique aspect of my criticism, I believe, is that it attacks an Achilles' heel. It looks at the fundamental reason for (that is, why) scientists put forth the Expanding Universe theory in the first place — historically. Let's go back. Examine the origin.

An Alternative World

I would now like to shift focus. I want to tie-in to a quite fascinating world. Something happened — a strange transformation of our sensibilities and thought. That shift has been around and available to us for a good while — about a hundred years. This new movement was an unexpected bonanza of creativity. It opened our eyes to wonderful, yet frightening and horrifying new things.

It happened during the most productive period of the amazing Albert Einstein's career. I suspect that this concurrence was not mere coincidence. I think there was a relationship among the thoughts of those living then. In the German language, we'd say the Zeitgeist (spirit of the times) changed.

At the beginning of the 20th century, the cubist movement pushed visual art into an entirely new direction. Soon after cubism came Dada. Our seeing sense of reality found itself warped and dissected into odd, cube-shaped blocks. And we soon had to cope with a flurry of absurdist, off-beat, surreal, kaleidoscopic and abstract images.

Likewise in the field of music, we began to hear atonal, discordant, maybe harsh, experimental sounds.

In turn, the psychological sciences peered behind hidden mental curtains. Freud and others boldly threw back veils. Those shades had kept shut secrets about our inner ids, egos and superegos, emotions, and internal experiences of sex, death and reality. This scared people because of its raw nature and intimate assertions.

A harsh, dictatorial communism plunged forth in Russia. Stalin took the lead. This bold revision of politics pulled the rug out from under current theories of economic privileges, commerce, value and prestige among the elites.

End-of-the-world scenarios seem to have become more popular in religious circles—perhaps as a result of the stressful novelty, cognitive dissonance and apple-cart upsets.

Humpty Dumpty fell off his wall.

So, what exactly can we call this different or "alternative" style? It was *a breakdown of time-worn, traditionalist habits of thinking*. Our slang for it now might be that it would "blow your mind." Some spirit, something, and more than one someone forged new tools for thought. A fresh view—another attitude—opened the way forward, for bad or good.

Thanks (or no thanks) to that "sea change" and paradigm shift, we have been freely considering new things for some time. I'm calling one of them a "bent universe."

Big Bang

What is the Big Bang? It says the universe is expanding because of a gigantic explosion at the beginning of time.

Parts One, Two and Three

I found it convenient to dive deeply down into three aspects of the universe. I will include:

- a booklet discussion about gravity and the Big Bang

- a short essay on irregular, non-linear forms

- a slightly longer booklet that discusses the mystery of time and how time can be elastic.

In each of the parts, we'll investigate an alternative way of looking at space, time, material substance, weight, motion and structure. That alternative pathway runs like a unifying theme meandering and warping through three parts of a "bent universe" discussion.

First, we will rise to a challenge of the Big Bang theory. How did we come to believe in this theory? What was its genesis? Is it true? Why or why not?

Next, in a casual and humorous manner, we'll look at a suggestion—that straight lines and boxes get in the way of scientific understanding. We'll try to decide if squares and rectangles make good engineering and design choices. Why do we like straightness?

Finally, let's try to pin down and tackle the history and nature of time. Can time and space be stretched out like a rubber band?

I wrote these three pieces at different periods of my life. Yet they all hang together with a similar mindset.

A Comment on Style

I don't expect to win any points for style because I know my somewhat whimsical or irreverent expression and attitude may rub some readers the wrong way. Tone and taste are personal. So I'd like to ask you, the reader, to be patient. Especially if you're knowledgeable in science, I ask you to try my way of talking about this. Would you do a quick once-through — by using a visual scan or a flip-through, or by moving briskly along with some sort of speed-reading? Thanks for your tolerance.

A Personal History Note

A quiet rebel looks into space. That's me. It's a big part of my story — of who I am. I am requesting that you share with me as we think about an idea I've worked on all my life. I sometimes wonder if I'm trying to honor someone and carry on for him — my older brother. He died tragically. I was 2 ½ years old. My mother wrote poignantly about Jack's curiosity. (Prideaux)

```
Jackie was almost six. He lived
on an Oregon farm. There was room
enough for the children to run
```

and play… But Jackie's world was much wider than his five-acre farm home. "Look, Jeanie—oh look, look, look! Mother, come see—", the night the sky was full of falling stars. After a trip to the beach, he puzzled, "What makes the waves keep coming in all the time and all the time and all the time?" … "How far down do you have to dig to get to the hot stuff?" … "If there are people on the moon, does our world look like a star to them?" … "Why does water run in through a hole in the bottom of a boat?" … "How can the radio waves get through the wall?"

…He went to bed one night with a headache. By the next night he was gone. There are no flowers on his grave… but a living memorial could pay tribute. Some flowers will never die.

There isn't anyone you couldn't love once you've heard their story. (Kownacki)

Once upon a time a boy peered at an ageless, lunar orb. Would you share that romantic boy's wide-eyed story?

Bibliography

Davies, Robertson. http://www.brainyquote.com/quotes/quotes/r/robertsond141553.html

Goldsmith, Donald. *The Runaway Universe,* Perseus Books, 2000, p. 6

Goldsmith, pp. 113–114

Kownacki, Sister Mary Lou. The *Reader's Digest* magazine, February 2015, p. 1.

Prideaux, Beth. Excerpted from "Flowers for Jackie," a personal essay submitted to Heifers for Relief for publication around 1947.

PART ONE

No Big Bang

There is an unquenchable and irresistible thirst of the soul that demands an explanation of the world in which it finds itself. (Singer)

Introduction

Who hasn't heard of it—the Big Bang? An overwhelming percentage of the astronomical and physics community has accepted and believes in the so-called Big Bang theory. Although I am an amateur thinker or theorist I have spent a good deal of time—many decades—delving into the question.

I see a flaw in the theory. I have tried to write a brief, critical, but constructive essay in common language.

I try to offer an entertaining, alternative idea. Hopefully, my casual style won't harm the value of my idea or interfere with anyone looking at it seriously.

I intend to reach, as my primary reader, the professional or expert scientist who has an open mind. However, I would also like to challenge all lay persons who are curious but may not know what some of the technical language means. Those non-expert folks, like me, can grasp some general concepts and get the big picture without needing to understand all the technical terms.

I discuss the effects of intergalactic gravity and propose a new concept "p." That "p" is simply an organizing constant: it mathematically suggests how the universe holds together, why all things move, how the cosmos is "bent," and how the supposed evidence may not support the Big Bang theory.

"P" may offer a hint to why many have misunderstood the so-called evidence that led to the theory in the first place. That evidence is referred to as the red shift. That questionable evidence, I argue, does not prove a one-time beginning of the universe and all things (the Big Bang). I explain why I think the universe is not, in fact, expanding.

If you remain skeptical after reading this introduction, I ask you to take a bit of time to read my short presentation. Then, I only request that you ponder "Maybe..." rather than reject this immediately with a resounding "No!" Would you give the idea a fair shake? Thank you.

No Big Bang

Was there a Big Bang? I don't think so — it simply did not happen. Maybe you believe it did, but I do not. I'll tell you why.

First off, it's only a theory. Big Bang is an idea that tries to explain something about our universe. Of course, we have to admit no one *knows* if it's true or not. That would be ridiculous to claim (if you said that you know — or someone else knows — *for sure*).

A theory is always subject to testing against facts and evidence to weigh its merits — its truth or falseness and its usefulness or not. That theory which, over the course of time, withstands all challenges is worth its weight in gold.

If no one brings forth any challenges, objections and doubts, then a theory remains untested. And so that theory left untested cannot be held to be valid. It's only tentative. It does not meet the "gold" standard of tried and true. It suffers from lack of exposure to adverse and alternative criticism.

So why is there a Big Bang theory? What happened to make people think of it?

Evidence

It all goes back to one lone and single piece of evidence. The entire idea depends on that singular evidence that is about 100 years old, give or take a few years. And if you remove that so-called "evidence" then you also will have to take away the theory which depends upon and tries to explain the so-called evidence.

I want to offer you something new, proposing that the supposed evidence can be better explained by another theory.

So what is the evidence I'm talking about?

It's called a "red shift." That simply means that light (such as a star) moving away appears redder, and a light moving toward us appears bluer. It's kind of like how a siren or a train whistle gets lower in pitch or higher in pitch. If a train is coming toward you or going away from you, the sound of its whistle seems to go "up" and "down." The sound doesn't stay steady and level.

In a literal way, the sound waves that reach your ears are heard *as if they are bent.* (Not your ears — the sound waves.) Actually, the waves are not bent. But to our ears it's *as if they are* bent. The extra speed of

the train, added to the speed of the sound waves, "pushes" or "pulls" the waves to a higher or lower tone or pitch (apparently, that is, not actually).

The actual sound heard by the engineer (who is traveling at the same speed as the sound) stays the same. It's only the *observer* watching the train go by who hears a "bent" sound. This is called a Doppler effect.

In much the same way, astronomers discovered that the light of distant stars seems to be "pushed" toward us (or "pulled" away). That light measurably shifts its color, toward or away from the red end of the electromagnetic spectrum (spectrum = like a rainbow or prism).

Think of a particular light hue in a rainbow as being *less* "red" in shade (more of an "orange" appearance) or *more* "red" in shade and you'll have the concept. Red light shading or bending (less or more intensely red) is like sound warping (shading) of a train whistle. So goes the "red shift" explanation.

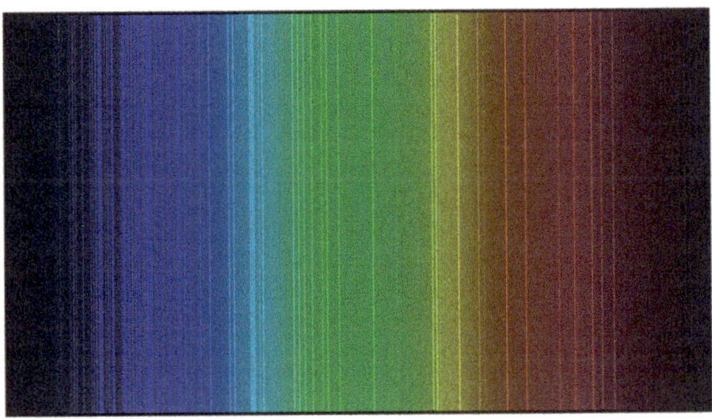

Keep in mind the star's light is either more red or less red, to our eyes, than the actual shade of color a star really is.

So what does this have to do with the claim of a Big Bang beginning?

Explanation

Let's look a little deeper now.

Astronomers discovered something very puzzling. They found a direct relationship between the red shift and the distance of a star from us.

The evidence of "red-shifting" when compared to the distance of the star from us (the observers) seemed to be a mystery, a curiosity, an important clue. We want to know what the galaxies are doing—understand space. And like all good mysteries, this intrigued people who thought about it. How can it make sense? What's the reason for it?

Astronomers decided it had to be *caused* by something. And the cause they came up with is their theory that our universe is *expanding*. The primary one who receives credit for establishing this is Edwin Hubble, who made up a "law" called, of course, "Hubble's law."

The expanding-universe theorists invented an analogy to help explain what that would look like. It's like a loaf of raisin bread puffing up and out in all directions as you bake it. As the whole loaf increases in size, the distance of each raisin from every other raisin increases also. We are supposed to be one of the raisins — so this idea says.

And from there — it's only one small step to say that the reason for the supposed or the possible *expanding* (of the universe) is that there has to be something to cause that expansion. How about a Big Bang? Did some primeval explosion kick it off way back when — at the beginning of time?

Now, there's a dramatic theory for you!

It hangs together. It's a nifty idea.

But hold on a second! What if we can come up with *another* reason that also makes sense (as to why the red shift happens)?

Instead of detouring into *how* the universe began (*if* it did; maybe it was always here), let's go right back to the so-called evidence.

All right then, why does a red shift happen? What's going on here?

Framing the big question

Since this is obviously a controversial point, would you allow me to digress a bit? I want to look back into prehistoric times, and I hope to try to build a case for why we need to challenge a Big Bang.

Most of us are familiar with early humans and civilizations. In the ancient and way-olden times, for countless thousands of years, people thought the sky and heavens were related directly and personally to humans. That was long before Galileo, Copernicus and others, and before so-called "modern science" came along.

People thought about themselves and about this planet (our home). They said: "we are the center of all existence." Call this view "anthro-geo-centrism." Because many believed that the Earth was flat and the sun goes around the Earth, someone could fall off the face of the Earth if that one (a mariner) sailed too far out in the ocean.

Some cultures that were non-European had other views. For example, Polynesians knew by experience that they could discover new islands in the Pacific Ocean if they just kept on going.

Greek thinking got the ball rolling, so to speak. But it wasn't able to get us to the moon and planets.

When newer science got a running jump on the older beliefs, the more modern view said we should try to think of our solar system and Earth as part of a "heliocentric" model (the planets go around the sun). That gave rise to modern astronomy and space travel and all the rest. This history of thought is well covered and most people know about it. So far, so good.

But now, maybe it's time for a new point of view: let's call this new, different perspective a "pan-centric" model. I want to say we could have a "pan-centric" reference point from which to look at the red shift.

Relativity and gravity

Before we delve into this "pan-centrism," we should bring into the picture something first identified by the famous author of Relativity.

Mr. Einstein said that our universe as a whole is curved/bent/warped. There are no straight lines…no parallel lines, and also no squares, boxes, rectangles or absolutely straight Cartesian graph reference lines on a galactic scale. This seems odd to us. How can that be?

It's gravity. Gravity tugs, pulls and pushes everything. It actually bends straight lines and makes them curved and crooked, even wavy. A measuring stick or ruler is really ever-so-slightly bent by gravity. A "straight" line from Earth to sun cannot exist. There is too much gravity: it prevents such an orderly, tidy, even and perfect line.

Gravity does the same thing to light. Almost a hundred years ago, during an eclipse of the sun, astronomers proved this bending did happen. They were looking at a star as it peeked out from behind (beside) the sun. Because the sun was darkened by the eclipse, the star appeared visible and measurable to them.

This proved that Einstein was correct. Gravity bends the starlight, just as the motion of a train bends a train's whistle.

The sun's immense gravity actually bent the star's light and the star appeared displaced from where it should have been.

Another analogy here is the displaced coin illusion, as seen in a glass of water. You can see a "double-image"—apparently. Light doesn't always travel in a straight line.

Should this gravity-bending be the same no matter where you are in the universe? Yes. One should assume the same physical laws that work here in our solar system also work everywhere.

Gravity accumulation

Clusters of "mega-gravity," "micro-gravity" and medium gravity make stellar and interstellar space a quite chunky, variable and swirly environment in which to live.

I think there must be an *accumulation* of all this measurable gravity at any given place. Accumulation should also be happening everywhere, from *any* vantage point.

Let's go back to the idea of "pan-centrism." Pan means "everywhere." We can grasp the idea that there is no **one** center of the universe. It's as if an apparent "center" is but temporary: only that point from which you take reference. In other words, it could be **any**-where. So gravity's bending is going on all over. It's universal — so to speak.

The cumulative effect (the amount) of that bending might be thought of as a constant "p." (Let's call it "p" for "pan.") Assume there is a "p" value that holds constant no matter where one is in the universe. Think of it as the "average gravity" per location of all the mass that exists. That average force exerts and applies its effect at any point in time and space (specific location) in the universe.

Now, if you send a star's light off rambling through the universe, it should be **bent** by the warping effect of gravity. And the amount of the warping should be predictably dependent on the distance of the star from the observer (because of "p").

So if "R" = red shift, and

"p" = the gravity constant, and

"d" = distance, then

$$R = pd.$$

The greater the distance: the greater the bending. The amount of red shift will be directly proportional to the distance of the light object from the observer—from any point in the universe. If you increase "d" in the equation, then "R" increases also—in a multiplier amount.

This math adds new input into Hubble's formula. Instead of speed/velocity (or acceleration), we put in gravity.

It's important to distinguish this idea from other similar ideas such as "gravitational lensing," which involves galaxies. Galaxies' lensed light resembles a single star's light in the famous 1919 experiment.

It's also important to distinguish it from what's been called "gravitational red shifting" or "Einstein shift."

In gravitational red shifting, light goes from a local region of higher gravity to a region of lower gravity. The reverse is "blue shifting," when light goes from a lower to a higher gravity region.

The idea of a "pan" constant is that we hypothesize the average gravity per point of *the entire* universe (not red or blue shifting in a *local* region/area with its local gravity variations, greater and lesser).

Illustration of how red shift works

Image you are looking at a star. Pretend that your star is moving away from you, in the same way a train moves away from you down a train track. You may explain the Doppler effect (measurable "red shift" of the star) by saying that the reddish shift is caused by the universe expanding and accelerating. This idea is similar to how a train whistle's sound is "bent" lower because it moves away from you. The red train symbol may be a helpful way to visualize this.

Big Bang

Imagine you are looking at the same star. Now pretend that your star is not moving away from you like a train does. Imagine that the star is shooting a "gravity ray gun" at you and you are shooting a "gravity ray gun" back at the star. You explain the measurable "red shift" of the star by saying that the red is caused by gravity, not by an expanding, accelerating universe. Accumulated total gravity makes the light shift toward the red end of the spectrum. The red ray gun may be a helpful way to visualize this.

No Big Bang

Back to square one

So, what happens to the expanding universe? Well, maybe it's just not expanding. Who knows? Who can say? The truth is there may not be the evidence for expanding that we thought we had.

For me, it's what we know we **don't** know that's fascinating.

If you take the so-called red shift "evidence" and subject it to this new consideration you'll get back to "square one." We must begin anew. We have lost the expansion "evidence" that we thought we had, with which to explain the origin of the universe.

Is that so bad?

I don't think so. All it means is that the field of cosmology and origins would now be wide-open to *other* theories.

So do we have to forget the supposed red shift "origin evidence" because it no longer is useful: i.e., it doesn't support Big Bang? OK—maybe so.

What about the curiosity we have regarding how this awesome universe got here? Were we created? Why are we here? Is there a God? What will happen to us? These big questions have been hanging around for quite a while.

Even though some 20th-century astronomers thought a "steady-state" universe made sense—as compared to a one-time creation—I don't feel we have to choose either-or (Big Bang or Steady State/ static state).

I rest my case here. It's enough to imagine and describe how all that gravity "works." Do we need a cosmological constant, as Einstein proposed? Do we need a unified field theorem? As for ultimate causes and such, I'll take a pass for now.

What's with all that gravity out there spinning us round and round? We know some of the cosmic swirls and eddies and surges occur in patterns that we can pick out of a chaotic background—obviously. We already have made a lot of sense out of what happens in our own vicinity or neighborhood (near-space). And we see markers such as supernovae that give us clues about the far, far and farther reaches beyond close galaxies.

What more regularities can we ascertain? If I were a practicing astronomer (I am not), I'd focus on the center of the Milky Way galaxy. Who knows where the next great idea will come from?

It's time for new thinking, in my humble opinion.

A critique

20th- and 21st-century physics is and has been handicapped by the accelerating popularity of ideas that appear to me increasingly nonsensical, strange and bizarre. Rather than **start** with Big Bang and theorize such ideas as "dark matter" or "black holes" or "alternate universes" or "string theory" or "supersymmetry," let's be willing to migrate away from such odd but sometimes appealing and impressive notions that constitute cosmology now. Things often seem just plain kooky out there. Could we take some of these things back to the drawing-board for redesigning?

It seems like the greater the gobble-de-gook, and the more fantastic the idea, and also the greater the reliance on technical jargon — the more impressed are many who settle for a status-quo view. It's been easy to go along with the Conventional Wisdom of the Big Bang, because of a bandwagon effect. And what other choices have been put out there?

In *Farewell to Reality*, Jim Baggott argued something very similar. To paraphrase: "We know more about the physical world than at any other time in history, but we understand much less." He went on to say, "physics long ago lost its grip on the real world we experience." (Baggott)

Why must we make the vast, wondrous universe so tortured and strange? I don't believe it has to be so out-of-reach for us ordinary mortals. I've tried to KISS (keep it simple, stupid). Will this succeed or not?

If you will excuse the expression, it seems to me our thinking is getting too constipated. We're stuck. We gaze at our own navel (at an astrophysical body of mistaken interpretation or *faux* understanding, that is). I think we simply got it wrong.

We got off-track

I believe those who chase after "proofs" of the Big Bang are chasing after mirages. If a theory is wrong to begin with, then no seeking after evidence can make up for the deficiency of the original concept. More evidence can't help to prove a wrong theory.

It may be time to throw it out the window!

It's been argued that success in science comes from asking the right questions. Perhaps my question will end up as a dud. Maybe someone will ask a follow-up question better than mine. Who knows?

I believe our science will be saved, and science can advance if we stay open-minded enough to consider new angles. Let's get un-stuck.

Arthur C. Clarke was fond of saying that truth is stranger than fiction. One rule of science holds that a simpler explanation is often to be preferred over a more fanciful, elaborate, complicated or difficult one. Plain is good.

Epilogue

Comprende? Do you understand?

Do you find it difficult to make sense out of the original, hundred-year-old evidence called "red shift" and its relationship to the Big Bang? If so, my alternative explanation (there was no Big Bang) will also seem hard to fathom. This stuff depends on the concept of relativity (as Mr. Einstein explained it) and that one is tough to wrap one's mind around.

I'd like to offer some simpler ways of understanding. First—an analogy about a dress. Next—an elevator analogy. Finally—some short summaries.

Early in the year 2015, a fascinating phenomenon went viral on the world wide web.

> In the beginning there was a dress. This dress was photographed and put on the Internet, like many dresses before it. That's where things got

> messy. This particular dress was particularly confusing because to some it appears blue and black, while to others, it is clearly white and gold. What began was a debate for the ages. It implicated God and man and the cosmos. (Slate.com)

If you were one of those who got pulled into this dress color question, as I was, you know the feeling of having your grasp of reality undermined. How can it be? What am I seeing? I can't doubt the truth of my own eyes, can I?

Well, yes.

I think I now understand how those dress colors could "flip." A varied presentation of light and color played a psychological trick on us. We saw one thing. Then we did a flip-flop. We saw another thing. We actually saw two distinct and different aspects of the same thing. It makes one think of the saying, "See things in a new light."

I offer this widespread illusion as an example of how minds choose to make sense of reality.

You may see I am saying that in a "bent" universe like ours, gravity and acceleration appear to be merely "flip" sides of the same thing.

Another analogy

Do you know why you can't tell the difference between acceleration and gravity? It's hard to get it.

If you increase speed in your car (let's say a souped-up race car), the sensation will be the same as feeling the force of gravity. As another example, we know the terms 2G (twice gravity), 4G, 7G and so on. As a rocket is launched and accelerates into orbit around the earth, enough force must be generated by fuel burning to bring the rocket's speed up enough to remain in orbit without falling back to the earth.

Now let's say you are in an elevator, in a very tall building. But you don't know where you are. Its compartment or room is tightly enclosed with no windows.

The elevator is accelerating toward the ground. You will feel lighter than normal gravity. You won't "weigh" as much (you feel) as you do standing on the ground. Let's say you briefly feel you weigh only one-sixth what you normally weigh, as you are plummeting downward.

Now imagine that instead of being Earth-bound you are sitting in that same elevator car on the surface of the moon. You sit there, not moving. Again, you weigh in (feel lighter) at one-sixth your terrestrial weight.

In both cases, you would have an identical sensation. In the one case, you're accelerating. In the other case, gravity (on the moon) causes the sensation. You cannot tell what the cause and effect is merely from the sensation evidence. Only if you have other clues to go by will you be able to determine the cause.

So what's with that dress color — black and blue, or gold and white? Is it flip — or flop? Heads or tails?

In our "bent" universe — gravity and acceleration appear to be merely "flip" sides of the same thing.

Summaries of No Big Bang

Summary 1

How often does an idea seem provocative and grab your attention? Maybe this one will.

"It doesn't pass the smell test." So says the author of a new book. A bold, new approach challenges a popular theory — a concept we all know as the Big Bang. Ka-boom! Will this book have any credibility among scientists? To do so, it has to bring in some technical language (astronomy). But it uses everyday words as much as possible.

Summary 2

No Big Bang—A three-part look at a bent universe is both a light-hearted, yet also serious, probe into the depths of the simple question: What are we? It takes a decidedly alternative look at space and time. It's about an age-old puzzle, the wonders of the cosmos, and relativity.

This work divides into three parts or angles:

- a section that brings gravity into a critical look at how the Big Bang has been justified

- a short essay on irregular and non-linear shapes

- a slightly longer section that discusses the mystery of time

Summary 3

This book dramatizes an alternative way at looking at the universe. Instead of looking at the boxy shapes and straight lines we all know so well, the author invites us into a "bent" or curved way of thinking. Somewhat like a simple band with a twist in it (known as a Moebius strip), our physical existence may fold in on itself at the cosmic scale. That's one thing Einstein's relativity is about. As a consequence of this, our universe may not be expanding after all — so no Big Bang.

Summary 4

If you take away the so-called starlight "red shift" evidence, you knock out the underpinning for the Big Bang theory. How can you take away that evidence? Argue it's caused by gravity. We know that gravity bends and warps our entire universe. Even time is not absolute. The straight lines we've drawn so long — our queues and boxes — get skewed by gravity, too. The cumulative effect of gravity could shift light toward the red side of the spectrum.

Summary 5

Some scientists lately have been wondering if Einstein's relativity got it wrong. But a new author now says maybe Edwin Hubble was wrong, not Einstein. The so-called Big Bang theory comes from evidence Hubble collected. Perhaps we should re-examine Hubble's information (the "red shift") — not the facts *per se* — but his (Hubble's) ***interpretation*** of the facts.

Bibliography

Baggott, Jim, *Farewell to Reality,* Pegasus Books, 2013, NY, pp. x, xi

Singer, Charles. *A Short History of Scientific Ideas to 1900,* p. 1, Oxford University Press, 1959, published in London, Oxford, New York

Slate.com. http://www.slate.com/blogs/the_slatest/2015/02/26/the_great_blue_and_black_versus_white_and_gold_dress_debate.html

PART TWO

Queue and Cube

Our lines, squares and primitive reckoning

Introduction

This short essay is cheeky and light. My point is that we may be unwitting prisoners of form, habit, and architectural fashion—more than we realize. Our urbanized and "built" environments haven't happened by pure chance. They have been chosen, selected, crafted. Our ancestors long ago invented forms we still prefer, such as ninety-degree angles.

But we do not need to continue our common, everyday acceptance of "square" and linear designs and widespread patterns that surround us. Being more aware of shapes can make our thinking more fluid, less rigid and more open. We can appreciate an irregular esthetic. We could better notice and admire the unpredictability, oddity and variety of nature's patterns. We could go forward with a kind of "back to nature" movement in honor of beauty, but make it an avant-garde movement that's based on modern, up-to-date math and geometry.

As I mentioned at the beginning of this book, an astounding shift of esthetics and sensibility took place in the late 19th and early 20th century. I think of the arts (cubism, surrealism, jazz, etc.) as leading the way. If we embrace a new kind of thinking, a "bent" universe begins to make more sense.

Straight down the line

Line 'em up? All in a row! Then square 'em off, for a great show. Set angles right. That's the way! Look smart! All in order! Tidy up and tuck it.

Now straight ——————————————⟶ down the line. That's the ticket!

How many of us look at life that way? Who doesn't love the straight, crisp line? Who doesn't love the neat box, or corners that make exact right or left angles? A 'T'? An 'L' shape? —yes, sir. Make those edges even. Box all complications nicely in, so they won't get out. How many of us get thrown for a loop when we try to handle odd angles, strange curves, irregular shapes, curly-Qs, unpredictable burps, twist or torque?

Much of our human heritage extols the Ts, Ls, and squares of the world. Some beliefs actually hold that a moral value can exist in such symbolism.

> The shape of his tiles, a perfect square,
> has been used by many cultures and,
> in Islam, is a symbol of paradise.
> (Hought)

Here and now, ***I declare war on the queue and cube!*** I throw down the gauntlet. I want more bent lines. Irregular curves. Helixes. Parabolas. Curlies. Wiggles and Squiggles. Asymmetry. Anomaly. ***Like the flight of a butterfly, or the meander of a stream.***

I call out for a new day in numbers — better stepping stones for common reckoning. Who will join me?

History of this

Look around you. Except for man-made, crafted structures, ***what do you see in nature?*** Any squares? Rectangles? Straight lines? Any right angles? I doubt it. Nature does not give us much of that. We see it only very rarely. The closest, dominating, natural (non-artificial) feature I can think of that seems to follow this pattern — is the horizon (an unobstructed view). And Earth's surface is ***actually*** curved, as we now know.

A Renegade Gardener, Don Engebretson said, "There are no straight lines in nature." He went on to advise, "Let the curves of the natural landscape be your inspiration." (Engebretson)

So where might we find in nature — if anywhere — a ***model of the right angle?*** How about this? Compare a "flat ground surface" with "straight up." If you assume the Earth is flat, then gaze straight up in the sky/heavens. That makes a 90-degree angle — line of sight. Your standing body forms 90 degrees compared with the flat ground. Well—almost. (Again, if the Earth were truly flat.)

When time began (hmmm, close to that ancient time anyway), great primitive priests and priestesses, shamans and esoteric thinkers reigned. These folks kept secrets and guarded sacred knowledge. They found out mysteries in stars, skies, seasons. One crafted a square. Voila! They began to proudly show off their new "square art forms."

Next, after hundreds of years of trial and error, they invented a way to measure a-squared + b-squared = c-squared. Presto!...they made a right angle, using theory!

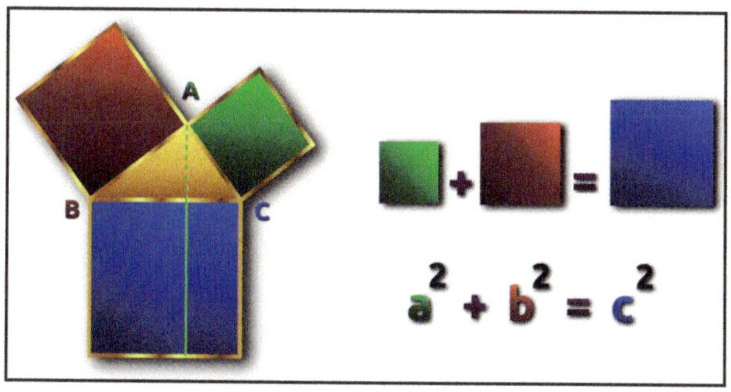

These ancient ruling class experts designed and built platforms. They explained (to initiates) how to practice mathematical, ritual, heavenly arts and crafts—on perfectly rectangular ziggurats, pyramids, altars and *heaaus.* They taught regular motion of moon, sun, stars and heavenly bodies. By measuring land, they mastered humans and organized labor. Thus they ruled. They civilized and settled worlds. From South Pacific to Mayan temple lands, and from Mesopotamia to China, rectangular structures sprang into lively, arcane being! The educated elite held new keys to earthly kingdoms.

Up to the time of Rene Descartes and for some time later, educated and privileged classes dominated. They graphed, mapped and surveyed, often with some glee and abandon! And much self-righteous satisfaction. Builders followed carefully squared-up plans — we call ours blueprints — significant, straight-line sketches.

But gradually — whoops, what do we have here? Educated trendsetters ran smack up against a problem, a thought that had begun with the work of the ancient Greek, Euclid.

Just try this: *extend* the logic of lines, squares and 90-degree angles. If you take it out far, you find out…hmmm, drat!…our world's not flat after all — it's curved and rounded. Nor is the cosmos squared. The universe itself is bent and warped and folded! **Space and time don't line up right and neat.** The time-honored theory of parallel lines finally failed! Parallel lines don't stay that way forever. They finally disobey.

Gravity marched slowly, heavily and inexorably onto the stage. Some new actors dramatized this entirely new story about geometry, "ether," mass, force and bending.

Who, you might ask, did the terrible deed? ***Who killed the ancient, long-standing belief in 90-degree angles and parallels?*** Several 19th-century mathematicians, that's who: Bolyai, Lobachevsky and others. And of course—Albert Einstein. Bold theorists got rid of certain, absolute, rigid, pesky parallel lines once and for all!

After 2,000 or more years of old math, some pioneers had birthed a "new" math. We now talk of relativity, fractals, chaos theory, field theory and the uncertainty principle. We have analog computers that deal with gradations and non-boundaries, and set theory for numbers and objects.

We began to see popular outbreaks of a new esthetic early in the 20th century. Frank Lloyd Wright pushed architecture upside down, sideways and inside out. ***Asymmetry gained a solid, attractive foothold.*** Heisenberg helped (uncertainty). Nuclear physicists did too (unseen particles). And Buckminster Fuller (geodesic domes). Beatniks, hippies. Andy Warhol and the art world. Anti–"little boxes" songs, and politics. What next?

It's only a matter of time before the construction and building industry, popular and folk art, and common people begin to seriously accommodate a new art and sensibility. New measurements and patterns. Cubism is out. Long live fuzzy blip and

crazy tilt! It's a whole new world of warps and woof!
I call it — these spins — "psychological math."

Vote for funky forms and livelier lines!

Watch for a new signal in drifting smoke, as it meanders with the breeze.

God bless America!

The End

Bibliography

Engebretson, Don. @*Home magazine*, the *Portland Tribune*, October 2005, p. 9

Hought, Nancy. "Beauty as Common Ground," in the *Oregonian* newspaper, October 15, 2005, p. B1, discussing the work of artist and sculptor Kanaan Kanaan.

PART THREE

The Art of Time: Crafting our Chronic Cosmos — or — Could Time Be the Third Dimension?

When you sit with a nice girl for two hours, it seems like two minutes; when you sit on a hot stove for two minutes, it seems like two hours. That's relativity.

Albert Einstein

So the sun stopped in the heavens and stayed there for almost twenty-four hours! There had never been such a day before, and there has never been another since, when the Lord stopped the sun and moon—all because of the prayer of one man.

Joshua 10:13–14 (The Book)

Introduction

How real is time? Questions about time seem to have become increasingly put forth into the common consciousness. As one example, consider what Sean M. Carroll recently had to say.

> There is…a judicious middle position between insisting on the centrality of time and denying its existence. (Carroll)

I support ideas similar to this "middle" theory or belief that Mr. Carroll refers to.

Would it surprise the reader to know that time is controversial?

On one hand, "absolute" time seems to be based on solid, age-honored and very persuasive physical evidence. It seems that science and technology have proved the case. But on the other hand, a number of arguments are very strong against an absolutist strait-jacket, and in favor of diverse and varied time. This other view says we experience a sort of "squishy" time that's not set in stone.

After pondering the question, "What is time?" I discovered it is easier to say what time is not, than to say what time is. Part of my reason for writing is to pose some questions that have no easy answers.

I would like to begin my exploration with three radical propositions. I also present these in the form of three questions. Then, I will look at how experts answer the questions. What authoritative opinions make sense?

I will explain how we craft time — how we make and shape our understanding of time — and how we measure and pay attention to the time that is important to us.

Finally, I will try to see where the debate might take us. The elasticity and relativity of time pose challenges to our future understanding of it.

Chapter One
Three Propositions

1. Time doesn't stretch out in a line. (Or does it?)
Contrary to popular opinion, time is not a line. We do use the expression, "time line." But that's not accurate, because time doesn't march by like a file of orderly soldiers on parade, one foot after another, one at a time. And of course, neither is a line of marchers perfectly even and regular. A marching cadence will never keep a perfect beat as does a metronome.

We might arrange days in a row like this: 1, 2, 3, 4. We could speak of yesterday, today and tomorrow, as if these days were called "day one," "two," and "three." We would thus count the day after tomorrow as "day four."

On top of that, we say a time line is a *straight* line. It doesn't curve, squiggle, hiccup, or go in circles. Furthermore, we imagine every second (or minute or hour) along that straight line is just as long as every other second. All these little chunks of time are equal to each other. Duration is constant and continuous, we say. But this is all incorrect.

So to summarize this point: Is time composed of units? Are time units arranged one by one like shoppers lined up waiting in a grocery queue? Can we put time on a line; say it's linear? Many would answer Yes, and say, "Of course our time line is correct."

I say, No, we cannot. Time won't always line up right. And I'm not the only one who has made this point.

2. Time is not sequential. (Or is it?)

Here's another commonly accepted aspect of time. One thing follows another in time. Of course it's true, we think. Sequence seems even more self-evident than the idea of a time line.

We speak of an event as occurring at a point in time. One hour, day or month follows another hour, day or month, in sequence. So we think. Events must happen before, now (concurrent), or after. We ask: "When did it happen?" In our English grammar, we use the word "tense" to describe the when state of a verb. Thus we speak in past tense, present tense, and future tense.

Example: "The girl sleeps. When she awakens, she will eat."

And that's not all. We incorporate the idea of sequence with another apparently obvious truth—causation. The notion of cause-and-effect depends on this kind of thinking. One thing happens after another thing *and because of* that other thing. "The boy threw the stone."

In this example, the boy's throwing arm moves. This action comes first. The flung stone comes after. We can visualize his arm moving and a stone *then* flying away *because* it was thrown by his arm.

But this kind of sequence does not always work. Here are some startling thoughts to consider. Time can become irregular. Time can vary. Times overlap. Time may apparently stop. Time may even seem to go into reverse, such as the case in which a dying person's entire past life flashes before his (her) eyes. Different people have different tenses.

> The purpose of time is to prevent everything from happening at once. (Kennedy)

Is there an "arrow of time"? We have become very accustomed to the idea that time is always regular and constant, and that time moves in only one direction—in a straight line ahead. We call that "progress."

3. Mechanics don't measure time. (Or do they?)

Finally, we believe our clocks are accurate. They measure small discrete units of time — seconds, minutes and hours. Our calendars measure large discrete units of time — days, weeks, months, years and centuries. Our sand clocks (hourglasses) make use of sand, gravity and glass regularity (artificial construction). Our ancestors also used water clocks to try to capture time. Sundials have tracked sunlight for thousands of years. Our sundials make use of fixed construction, earth rotation and angles of shadows. These older methods pre-date our modern, metallic devices.

Modem technology has employed physics to peg an atomic clock at molecular level events.

> The duration of a 'second' is equal to
> 1,192,631,700 cycles of a frequency
> in the isotope cesium 133. (Levine,
> p. 27)

Modern technology uses astronomy to compare one very large natural phenomenon to another very small one. For example, we could compare revolution of the earth around the sun to radioactive deterioration of carbon (carbon dating). Thus we obtain a ratio in nature, of one time "unit" to another "unit."

But our modern mental and commercial dependence on mechanical time dates back less than one hundred fifty years. What? That recent? The reader may challenge that. You might think our sense of time has been the same as it is now for a very long while. But, no. It hasn't. Consider the following. This example was one of the first attempts to commercialize time.

> In 1871, the Pennsylvania Railroad declared Samuel P. Langley's Allegheny Observatory time their official standard. They contracted for $1,000 a year to receive time signals from the Observatory. (Levine, p. 66)

This is but one illustration of a huge move in the 19th century and early in the 20th to standardize and promote clock time. A new, public time sense quickly emerged. Timepieces became popular. Carrying a watch became a status symbol. Now, one might ask: what came before the widespread buying and selling of time devices, machines, signals and coordination?

The simple answer is: ***No such mechanical time existed.***

Or rather, it didn't matter. It was spotty and sporadic, and it was of very little consequence. Perhaps it would be fair to say time measuring devices did not make much impression on human life. Mechanical devices were novel, rare or crudely made. They had not "caught on." For a long time, they were irrelevant and unimportant to almost all human activity.

True, the idea of absolute time did exist. European scientists in ivory towers imagined mechanical time. A select, few people thought it was a good theory. A limited number of huge public clocks existed. The big clocks were viewed as curiosities. But most people ***couldn't have cared less,*** in regards to their struggle for existence. Daily life proceeded without "our" (modern) time.

Also, in a very significant sense mechanical time is not real. It's useful. But it's artificial and unnatural.

Chapter Two
What Is Time?

Might we overlook patterns?

So, how did these time controversies arise? What are we to make of these provocative questions?

To repeat: 1. How is it that time isn't a straight line? 2. And why isn't time always sequential? 3. And what does it mean to say machine time doesn't measure what we think it does?

Let's begin with: We live in a world of patterns. We rarely give these shapes or regular events a glance or a thought. They are second nature to us. Many patterns are artificial—designed by humans. In parts of cities, *most* patterns are artificial. If one becomes more *aware* of the neatness and appeal of flat surfaces, straight edges, boxes and 90-degree angles—what one could call "straight line" thinking—one sees those patterns in a different light.

They cease to be merely background impressions. They become cultural tools.

Time as we normally know it is one of those many "straight line" patterns surrounding us. And we mostly *like* straight lines. They appear somehow neat, orderly and clean. We usually like modern time. We see it as an ordinary, normal and natural part of our world. We assume absolute time is the only time that exists.

But a strict time line is *not* natural. If we dig into a bit of history and science, we can discover how we actually "construct" the time that seems so obvious and familiar. We design our patterns. The art of time has a storied history. In a very important sense, our modern time is real, but is also a figment of our imaginations. We *make* it.

What is time?

Time has many rich meanings and wide significance. Its common sense meaning is "duration" or "a point when..." But we immediately run into trouble if we try to go beyond casual and colloquial uses of the word. For a list of dictionary synonyms, definitions and associated ideas, please see Appendix 1. At the end of this part, I once more ask this same question — and try to give a better answer to it. Here now is a more probing look at a sometimes odd, wily or perplexing concept.

Does nature create time?

1. The body

Our human bodies and genes generate internal senses of time. We are aware of some of our anatomy. Examples: heart rate, breathing rhythms, and discharge of the menses.

Other bodily functions are hidden. We neither sense nor notice them. Bodies are governed by unseen sympathetic or parasympathetic nervous systems. Some inner workings are related to time, and some are not. Example: contraction of blood vessels.

Nature has provided many organisms with amazing biological clocks. Here's a story about that sense of timing.

A Swiss doctor, Auguste Forel, used to enjoy breakfasting on his terrace. In 1906, Forel made an observation that was to change the course of scientific history. Every morning, at exactly the same time, bees from a nearby hive would show up to sample the jam on his breakfast table. Even after Forel began to breakfast indoors, he would notice that the bees continued to show up on the terrace like clockwork, precisely at the accustomed time, foraging for the jam.

> Forel concluded that...the bees must
> have some kind of built-in memory
> of time. (Rifkin, p. 30)

Biologists have also found that "not all biological clocks are circadian." (Rifkin, p. 33) Living things respond to their external worlds as well as to internal clocks.

2. The heavens

From the point of view of the astronomer or anyone who gazes at stars in the sky, our strongest human time sense is the diurnal, or day-night awareness (Earth rotates once a day). Example: we sleep and wake.

We also possess a weaker sense of astronomic time—the seasonal (Earth revolves around the sun and is tipped or inclined on its axis). Examples: average air temperatures vary by season and our bodies adjust; leaves turn color and fall from trees; amounts of precipitation — snow or rain — vary by season.

Another weaker sense is lunar (moon revolves around the earth). Examples: ocean tides; female menses.

An unknown sense is the solar systematic (e.g., the gravity interactions and pulls of various planets as they revolve around the sun). Much of astrology believes "the stars" (sun, planets, moon) influence us in specific ways.

Another unknown sense is the galactic. Our sun and solar system apparently revolve around the Milky Way galaxy once every several hundred millions of years. There also could be some kind of extra-galactic influence.

Included among possible solar-system or galactic time senses may be pulses, rhythms, repetitions and other patterns we see coming from various locations in "outer space." We observe, track and record these radio and light sources and other frequency sources. Examples: sunspot cycle (every 11 years?) and the "solar wind" and auroras that accompany the sun storms; distant pulsars; interactive motions of the galaxies.

3. The sub-microscopic world

Events at micro-physics levels, such as atomic or quantum, may drive and operate certain time "clocks."

Some thoughts here:

(a) Such incredibly small, sub-atomic clocks would not appear to affect human consciousness. (b) Such micro-effects (tiny clocks) may themselves be influenced by larger, macro-effects such as the rotation and revolution of planets. (c) Tiny clocks would probably have been determined by the physical laws of matter, by the original creation of our planet, by Earth's resulting composition, and by later volcanic and geologic forces. Example: radioactive decay of uranium.

There are two kinds of time.

1. Objective.

We could speak of derived or "objective" time. In the mathematical sense, this is the time we measure. Although this clock time has become popular only recently, it continues to use the ancient counting system of 12/24/60. We consult our clocks, watches and other timepieces to monitor time. We refer to printed calendars. The first famous reference to this so-called "modern" time was the expression, "absolute time." It has also been called "Euclidean." We could give it the nickname: "number time." We should point out that objective time is artificial and arbitrary. Since Einstein, our ideas of objective time have been altered by relativity.

2. Psychological.

We could also speak of "psychological" time — perceived time, impressions, personal experience. It's that time which we sense, observe and intuit in the absence of any time measuring devices, other tracking technology, or information based on such devices. A psychological time experience is flowing and fluid, sometimes fleeting, and sometimes fickle. It's the eternal rhythm of common, daily life — of the people, by the people and for the people. It's natural time: plant and animal tempos. It's poetry: the moon's course, river's meander and weather's seasons. It's the guidepost of birth. It's the swing of death's scythe. It touches something deep within our beings.

Psychological time is the only time we had for countless thousands and millions of years. It's what humans did and felt prior to the rather recent invasion of tick-tocks, Greenwich time, daylight saving time, workplace punch clocks, digital displays, electronic numbers like "8:23" that blink, and the idea that time is money or that "time's a-wastin'."

How did time begin?

Let's set the stage with one of the Greek myths. In the beginning was chaos. Then Uranus (the heavens) married Gaia (the earth). One of their sons was Chronus (which means time; he also is known as Saturn). He and his brothers and sisters are referred to as the Titans. Chronus married his sister, Rhea. Together they gave birth, among many other children, to Zeus (Jupiter).

This is one of the earliest recorded explanations for the universe and how time came to be. But there are other creation stories, of course. Genesis tells one. And modern science has yet another theory.

Chapter Three
Objective Time

How did objective time begin?

Of the two kinds of time — *objective* and *psychological* — the one is much newer than the other. To find the first measurement of modern, *objective* time, we have to go back in history only a few hundred years. And we must begin with a look at the modern idea of *space*.

The story unfolds in Europe following the Middle Ages. The emerging rational mood in Europe led to the era of Enlightenment. The new mood produced giants of science and mathematics. One early pioneer was Rene Descartes. At this point, we discover an invention by Descartes — a *new math of space*.

The "Cartesian" system has given us the box, graph, grid or dimension model. It is well known to high school geometers. It introduces us to some crossed, perpendicular lines: the x axis, y axis and z axis. And there is a point of origin at 0,0,0. We have now identified height (h), width (w), and length (l). Similarly, we identify North, South, East and West on maps, and to these points on the compass we add the directions of up and down. Now assign numbers to represent distance along these lines. Then multiply the three dimensions of space (h times w times l) and you obtain volume. This is very basic math for engineers and students of geometry.

Note here that it's only one small step of logic to go from this clear concept of linear *space* dimensions to thinking of time also as a *line*. But we're getting ahead of our story.

Mathematicians have for centuries described our common **Cartesian** grid space as **"Euclidean."** That's because the geometry of Euclid can be handily drawn and mapped onto any surface (the grid). It's clear. It's easy to see. It's easy to work with. A piece of paper has two dimensions on which to sketch. For example, one could show two straight lines "AB" and "CD" parallel to each other in any quadrant in a two-dimensional grid (2D: x axis and y axis). The same can be shown in a three-dimensional grid (3D: x, y and z axes).

Although Descartes apparently didn't refer to **absolute** space, his new mathematical structure created the opportunity for someone else to name it. The dazzling new math gave birth to an entirely new idea of space. And from space came time.

Incidentally, this is where we first picked up our concept of *dimension.*

Who invented absolute time?

That intellectual giant of mechanical physics — Sir Isaac Newton — built upon the Cartesian system. Mr. Newton created our ideas of absolute linear time and absolute linear space. "Beyond anything that happens, there is the measured flow of time; and beyond any place that it happens, there is the unchanging vault of space." (Berlinski, p. 105)

In Newton's own words, we have:

> **The law of absolute time.** Absolute, true, and mathematical time, of itself and from its own nature, flows equably without relation to anything external. (Berlinski, p. 105)

Newton's view of space mirrored this description of time. For him, both time and space existed as fact irrespective of any other considerations. He laid them down as axioms, fundamental principles, true by definition and true *a priori.*

For thoughts on how so-called absolute space dimensions are arbitrary and constructed, please see Appendix 2.

Who invented non-absolute space?

Something happened in the 19th century to throw

this 3D system (Cartesian grid) off kilter. Very few people today know the story. It's amazing. It's perplexing. It's a math puzzle: Can you prove parallel lines never meet?

A great turning point in the history of thought came as mathematicians (Bolyai and Lobachevsky were two) showed Euclidean geometry to be incorrect. Two parallel lines do *not* extend to infinity (according to non-Euclidean geometry).

One way to picture this is to extend straight, parallel lines along the surface of the earth. Go past the horizon. Then continue to extend the lines along the surface. They would go all the way around the planet. ***The lines circle. They don't remain straight. And the lines cross*** (if extended along the paths of greatest circumference of the sphere). But to grasp this new idea in a visual sense is easier than to prove the problem mathematically.

If one were to extend parallel lines for a long distance in space, they might come together.

> Because it is impossible in practice to measure how far apart the rays will be extended millions of miles, it is quite conceivable that man is living in a non-Euclidean universe. (Because intuition is developed from relatively limited observations, it is not to be

trusted in this regard.) (New Encyclopedia Britannica, p. 958)

Is there also such a thing as objective time that is non-absolute?

The answer is Yes. But in order to explain this answer, we have to take a quick tour through Einstein's theory of relativity.

General relativity brings in a fourth dimension

As Einstein showed, adding the ingredient or element of time (t) to the conventional 3D mix (height, width and length) makes possible what we call the "space/time continuum." Thus we have the notation: x, y, z, t. He took this idea from the German mathematician Hermann Minkowski.

Einstein invented a new way on a new scale to determine mass, velocity, gravity, energy and acceleration. He said these all exist in—and are part of—both space and time.

All the stuff of existence has mass and weight. And also, it moves and changes. And it moves within the context of, and as part of, a space-time continuum of four dimensions (4D), said Professor Einstein. Using the grid idea, we now write *"t, x, y, z"* for the four dimensions. Very simply put, this algebraic phrase would be read as ***"a location in space at*** *a point in time."* Or even more simply: *"There/now"* or *"There/then."*

Energy, mass and motion actually **define** that continuum. Each aspect of this physics is dependent on and **defined by** the other aspects. Each aspect exists **relative** *to* every other aspect. Each of the four dimensions is tied to all the other remaining dimensions.

Because of general relativity, time—like space—must also be "non-Euclidean." Why? I think it would be intuitively impossible to argue that space is non-Euclidean but time is Euclidean. I think it would be impossible to say a straight line in space can be bent by gravity but a straight "time line" can **not** be bent by gravity.

Time is bent by gravity

The general theory says that in an environment of heavier gravity (or when acceleration happens), time moves more slowly. In lighter gravity, time moves more quickly. If this is true, then why does time seem uniform to us? Why don't we notice variations? The reason is that for millions of years, we humans have occupied a very specific place, with about the same gravity everywhere (the surface of the Earth).

Four-D makes an argument against four, fixed dimensions

In a strange, apparently contradictory way, the concept of 4D can totally free us from dimensions. The notion of space-time continuum is not inextricably tied to three or to four dimensions, or to any one number. Mathematically, one could describe that space-time continuum *without* reference to dimensions. Or we could use fewer than or more than three or four dimensions. I think it might make more sense to do away with dimensions and talk about just *two elements* or two aspects of material existence (i.e., space and time). See also Appendix 2 about arbitrary dimensions.

Chapter Four
Psychological Time

What is psychological time?

Imagine you feel "lost." You go away from all sources you rely on for the correct time of day. No clocks. No smart phones. No digital displays. No mechanical or electrical devices. You're on a camping trip, a trek. You go to a wilderness reserve. Or you travel in a rural, remote area that is far from what we call "civilization." Perhaps you are floating at sea—on the great ocean, in a tiny raft or boat. Now you see countless stars at night.

Here you can sense the way our ancestral human beings evolved to "tell time."

For hundreds of thousands of years, humans *invented* time. Humans defined time each and every day. Each year we had to invent our timing anew as we listened to our hearts. We listened to the wind and saw the seasons. We listened for the sounds of predators that could eat us. We listened to each other. We listened to the drums of dance and song, and to human tempo. The cues of nature and the internal cues of our bodies told us when to act, and when to do something, or when to **not** do something.

That's the way human beings evolved to "tell time."

We get premonitions. We may get hints of this natural time—at odd moments. Slowly, over the course of many generations, various humans developed "time languages." We could identify key components of natural time. Farmers, agriculturalists and naturalists have sensitive vocabularies about rhythms and rivers of life.

One of the key components was psychological. How do we feel about time? How do we sense it? What do we think about it? What's important? Perceiving can turn into believing. Habit and culture reigned. A person's individual character might make time into a very singular event.

A wise woman I know has remarked often: "Isn't it strange how two people can be at the same place at the same time and have different memories of the same thing?" (Holling)

A Latin saying puts it this way: *De gustibus non est disputandum.* Taste is indisputable. One person's like is another's dislike. How can you account for that? You can't.

Your preference or your "taste" for time runs on a different schedule than does *my* "taste for time." Our inner or biological clocks can differ. It's all relative. Nothing is set in stone. One's timing can be "off" in regard to another's timing. We easily identify ourselves as "day larks" or "night owls" — referring to when we like to go to sleep at night and when we like to rise in the morning (earlier or later).

Our surroundings often dominate. How does time seem to elapse? Environments can shape and bend us into odd, unrecognizable forms. We react to what's nearby. Take an extreme example: a man's experience as he endured the Holocaust. Roman Kent survived a concentration camp. He recently stated:

> "A minute in Auschwitz was like an entire day, a day was like a year, and a month an eternity." (Kent)

We humans do seem to have some kind of biological clock. Some people can wake at a prescribed time with no pre-set alarm and no wake-up call. For them, sleep does not dull the internal, unconscious time sense. Perhaps we use subliminal perceptions as a kind of self-hypnosis. And surely many animals, especially migrating birds, can "tell time" from the signs and clues of nature.

What is timelessness?

Most of us have occasionally experienced an odd, strange sense of timelessness. (And I'm not referring to over-sleeping: getting up late with dizzy head and droopy eyes from a hangover.)

Did time once seem to stand still for you? Did you lose track or forget to notice time? Perhaps one gets caught up in the moment. Emotion can over-power logic. What we're doing may be fascinating — very involving. We "space-out." Or we get into a "groove" or "zone." We may pretend and play-act; get into a drama that seems real. We may feel lost or confused. For a while, time becomes unimportant, irrelevant.

We may recover from meditation or a "trance." Suddenly, we're back in "real time" or reality again.

One may have a spiritual experience: see, hear or know God or the Divine. One may be awestruck. We all have brief "aha" moments. Or one's eyes "bug out" or get "round as saucers." Or one's "jaw drops open."

This kind of experience, however, often does **not** seem strange at all—because it's very natural. In fact, it's probably more natural to live life **without** a smart phone or timepiece around one's wrist than it is to have that device and be a slave to it. Better to not obsess about a clock on the wall than to look at it often. Watching the days elapse by marking them on a calendar can slow down the passing of time.

One remembers an old adage: A watched pot never boils. It's sometimes a question of how attentive or **how fussy** we choose to be about time (i.e., our awareness of it). Perhaps we are required and obligated to set a mechanical alarm to wake us at a specific time in the morning.

During those special, unusual moments or periods when a sense of unreality and alternate consciousness sets in, time loses its hold on us. We may judge this is a good thing or a bad thing or neutral. It all depends on the situation.

But regardless of how we value the experience, objective time then **ceases to exist for that person or persons.** We have entered fully into **psychological time.**

Writers cite time warps

A mountaineer wrote of nostalgia:

> I have a strange sense at the moment
> that time is losing its attachment to
> reality, that it's no longer the mecha-
> nism that separates events but rather
> in its absence events are overlapping,
> even coexisting....I don't want to
> break this moment, to bring it back
> into time. (Ridgeway, p. 204)

Another writer looked back on the tumult of the
1960s:

> Dewitt remembered how for almost
> all of his life time had dragged...Then
> suddenly, as if gravity had loosened
> its grip, events began hurtling. It was
> like time had slipped its chain. Some-
> thing momentous was happening.
> Things in America were going too
> fast. Mach I, Mach II...Had some
> cosmic curtain been ripped open
> and we'd passed over to a hyped-up
> universe? (Hetzner, pp. 86–87)

Psychologist Robert Ornstein mentioned living
"in-time." It means: able to experience action as if it
were occurring in the infinite present.

One of the primary tasks of Zen is learning to experience the here and now so utterly that time appears to stand still....Tennis great Jimmy Connors has described moments where he felt he'd entered a "zone."...The ball would appear huge as it came over the net and seem suspended in slow motion. In this rarified air, Connors felt he had all the time in the world to decide how, when and where to hit the ball....Basketball players describe unexplainable occasions when everyone around them seems to move in slow motion. During these moments they report a feeling of being able to move around, between, and through their opponents at will. (Levine, pp. 33–34)

Poetic time

Joshua commanded the sun to stand still.

Was it a miracle? Or was it merely exaggerated or primitive language? What can explain the extraordinary event reported in the book of Joshua? What could be a correct interpretation? Modern astronomy tells us it is impossible for the sun to stop in the sky.

I believe both the writer and the Bible hero (the character) tapped into valid, psychological time. The literary image is certainly a vivid and poetic way to explain a very important, amazing military victory. We might not take it literally, but rather as metaphor. Its telling and its intent are certainly to extol God's power and to encourage believers and followers of Joshua.

Such claims speak volumes. To paraphrase a famous adage, we could note that one "word picture" such as this is worth a thousand words.

However, on another level of interpretation one could find a more literal truth in the Joshua story. From the perspective of someone who is not standing on the Earth, which rotates once daily, the sun *does* stand still in a day. For example, a moon-walker sees the sun appear to stay in the same general part of the lunar sky during a twenty-four hour period (Earth time). Thus that lunar or *extra-terrestrial or celestial* person could truly report the sun standing still for a day.

The magic of tenses

What are past, present and future?

Although in our modern life we may believe in secular and material reality, humanity carries a huge tote-bag of magic from the past. Many contemporary people still make use of superstitious practices and alternative styles of life. Our ancestors and early humans employed a well-developed magical world view for their spiritual well-being. Shamans, diviners and healers helped keep healthy and united the groups that depended up on them for vision and practical skills. Language and art reflected upon such mystery. Some of these pre-scientific ways have been preserved.

> The essence of magic is to exist in a state of consciousness where past and future seem interchangeable. Classical Hebrew, for example, has only two tenses: There is the present, and then there is another tense which barely distinguishes between past and future....A primitive sense of existence is suggested—one that would transgress our modern separation between the real and the imaginary. In such an ancient grammar, yesterday's events are not seen as facts which have already occurred so much as intimations of the future, that is omens received from a dream....To say, therefore, that you have done something which you have not yet done becomes the first and essential step in shaping the future. (Mailer, p. 569)

Chapter Five
Cultural Time

Is time the same from one culture to the next?

No. It is not. Just as two individuals vary, so do two societies vary. Each society or culture experiences psychological time with a peculiar character, in its own right. Vast gulfs divide human spaces, styles, relationships, perceptions, personalities, rules, histories, values, tempos, calendars and beliefs. Group differs from group.

I wonder if the following peoples posed time as linear:

- star-gazers amid the hanging gardens at Babylon

- Druids at Stonehenge

- Mayan sages in Central America

- devout followers of Mohammed in the 800s

- ancient Irish Christians

- palace sages in Great Zimbabwe

- early Chinese astronomers

- priests along the Nile in ancient Egypt

I would guess none of these persons who became experts saw time as a line.

One writer quoted a vivid encounter concerning time in our current worlds. He pointed to the time senses of some people of Trinidad and of a social researcher: James Jones, from the University of Rhode Island. It's the difference between a rush-rush urban style and a less hurried, more rural way of life. I've taken liberty to use slang here, only to make a point.

> Latecomers to appointments would greet his impatience with comments like: "Eh mon, what's your hurry, nuh? De sea ain goin' no place. Relax mon, a'm comin' to yuh just now." "So," as Jones puts it, "I wait." (Levine, pp. 81–100)

Many people look at time from the point of view of their important events. This has been called "event time."

Let's look at another example: Australian native peoples' idea of "dream time." (In the following report the native Australians are called Real People):

> Real People are aware of dream consciousness while awake....The tribal people do not dream at night unless they call in a dream. Sleep for them is a time for important rest and recovery of the body....They believe the reason Mutants [white or Caucasian people] dream at night is because in our society we are not allowed to dream during the day... (Morgan, p. 115)

This psychology differs sharply from the mental idea of lineal time. The author explained further about counting days. She experienced amorphous time with the Ancient Ones.

> There was no differentiation in days of the week while living with the tribe. Nor was there any way of knowing in which month we were living. It was

apparent that time was not an issue. (Morgan, p. 137)

Many Asian believers claim life is a cyclic and repeating process. They teach that the soul has re-births, passes through successive bodies, and experiences reincarnations (metempsychosis).

> Since metempsychosis presupposed a cyclic view of time, it was ruled out in religions conceiving of time as linear (e.g., Judaism, Christianity, Zoroastrianism). (Brandon, p. 439)

Hindu doctrine said that...

> ...liberation was to be pursued against the backcloth of cosmic processes of immense or infinite duration. (Brandon, p. 439)

> Cosmology is made up of recurring periods of creation [ebb and flow] until a new Brahma comes into existence. Each period is a "kalpa" or 4,320 million terrestrial years. (Brandon, p. 618)

Chinese Tao philosophers taught the relativity of both time and space. (Brandon, p. 618)

What is a week? Is it a cultural time frame?

The unit of time called a week is, without a doubt, culture-based. Your week may not be the same as my (seven-day) week.

Living is certainly easier when we track and organize our time. Most peoples of the world string a number of days together. They chunk days into larger groupings.

An extraordinary example of day-grouping comes from the French Revolution. New thinkers produced a unique calendar with 30-day months. Each month subdivided into three 10-day weeks. (Rifkin, p. 76)

Rifkin has pointed out: "Whether sacred or secular, every calendar expresses the essential politics of a culture." (Rifkin, p. 71)

Many week variations exist:

> The Incas had a ten-day week. Their neighbors, the Muysca of Bogota, had a three-day week. Some weeks are as long as sixteen days. Often the length of the week reflects cycles of activities, rather than the other way around....The Khasis of Assam... hold their market every eighth day. Being a practical people, they've

> made their week eight days long and named the days of the week after the places where the main markets occur. (Levine, p. 93)

The calendar to which we are accustomed in the United States makes use of a time-honored week with seven days. It [the seventh-day or "Sabbath"] has a history that goes back thousands of years:

> Among all of the Jewish time markings the Sabbath remains the most hallowed and the most honored observance....The Sabbath is a holy day. (Rifkin, p. 72)

Cultural controversy surrounds many yearly calendars

What is a year? As an example of differing years, students of European history may be familiar with a major time adjustment that re-calculated the yearly calendar then in use — instantly — by eleven days. In England, citizens changed their calendars from Julian or Old Style to Gregorian in 1752. For example, October 19 at once became October 30. This change met with resistance. (McCullough, p. 30)

Holy days are observed in unison

Holy days for most religions are fixed in advance. It

is a truism that if believers are to worship together, they must be in date unison for special observances. That is also true for cultural observances. These events bond and solidify members of a group.

One of the earliest types of calendar systems followed the moon. The Egyptians did this. But one Pharaoh abandoned moon time. He imposed a radically new, sun-based calendar. Akhenaten (died about 1334–6 B.C.E.) revered Aten as God. This created terrible controversy in Egypt. It caused so much resistance and furor that later Pharaohs chiseled out the name of the innovator. They re-wrote history and swept away disruptive sun time. Other cultures clung to a lunar calendar for millennia, and many still use it.

Following are examples of two lunar calendar cultures: the first is from history and the next is modern.

> [Hebrews] anciently went by the phases or appearance of the new moon...the new moons and festivals were then constantly settled by the Sanhedrin at Jerusalem. (Prideaux, p. 230)

...Islamic calendars are still primarily lunar rather than solar. (Fraknoi, p. 67)

The primary Essene monastery had its own calendar
We have learned, since 1947, much about the Essene community of Jews. These people existed before, contemporary to and directly after the time of Jesus. This new knowledge came by way of a mid-20th-century discovery. Ancient texts were found in Qumran caves near the Dead Sea. Scribes had protected sacred writings. A thriving colony of devout ascetics lived in a monastery built on cliffs overlooking the sea.

We know "orthodox" Jewish Essenes held their festivals and rites in great reverence. They believed God appointed calendar dates from the beginning of time, as in the Genesis story. God stipulated holy days and annual festivals. God gave these time-tables as part of the grand design of the universe — so they believed

Using a 364-day solar year, covenanters at Qumran organized weeks and Sabbaths into seven-day patterns — repetitions. This Qumran calendar also assigned Essene priests an on-duty schedule of dates within the solar year. Therefore, priests knew precisely when to conduct their rites and carry out responsibilities.

Qumran holy men conflicted with their competitor priests nearby in the capital of Jerusalem. Leaders of the Sanhedrin seemed to have used a lunar calendar. And, temple priests would have rejected dates based on solar time. In stark contrast, Essenes embraced a holy significance found in the beloved solar calendar.

> This strict community preserved its religious beliefs and practices, unpolluted by Jerusalem and its Hasmonean high priests...the Teacher of Righteousness revealed the proper interpretation of the Scriptures and the calendar... (Barnstone, p. 223)

In the Damascus Document, found among the Dead Sea Scrolls, Essenes proclaimed that God opened up before them certain times for celebration...

> His holy Sabbaths and his glorious festivals... (Barnstone, p. 226)

And, they declared how to punish

> ...every one who goes astray [because he] profanes the Sabbath and the feasts... (Barnstone, p. 232)

We also know the Essene calendar existed even before the Qumran community was established.

> This follows from its presence in earlier works such as the Astronomical Book of 1 Enoch and the book of Jubilees. (VanderKam, p. 61)

An Easter dispute made Christianity less Jewish.

Although I run the risk of dwelling overly on calendar matters, I'll present one more colorful example of a struggle concerning timing of a yearly date. I do this because the organization of yearly events is central to the culture of a people. Rifkin wrote succinctly:

> The Passover-Easter controversy is a graphic historical illustration of

the intimate relationship that exists between calendars and group identity. (Rifkin, p. 75)

A Christian ecumenical council was convened at Nicea in A.D. 325. It fixed a date for Easter.

> This calendrical change marks a major turning point in Church history. While the old guard was intent on maintaining apostolic tradition, the reformers were anxious to emancipate the Church from its Jewish parentage. (Rifkin, p. 74)

For a while, the Easter date debate unsettled believers who lived on the islands of Britain and Ireland:

> ...the Saxons came into this island (AD 449)...they cut off all communication with Rome...until the coming hither of Austin the Monk...about one hundred and fifty years after [thus]

> Whereas the Romanists observed the beginning of the Festival from the 15th day of the vernal moon to the 21st inclusive, according as the

> Sunday hapned within the compass of those days, the Britain and the Irish observed it from the 14th to the 20th. That is the Romanists laying it down for a principle in this case never to begin the Paschal Festival at the same time as the Jews…they were so zealously set this way that they would not hold Communion with those of the British and Irish Churches, that did otherwise, but looking on them as heretics… (Prideaux, pp. 238–239)

Is time ethno-centric?

Science has often discarded theories that once seemed self-evident in their cultural contexts.

Faced with new ideas from Copernicus to Darwin, traditional theories dropped off by the wayside. From the old Church to cultural secularism in the 20th and 21st centuries, new concepts led to changes in world views. New truths dethroned hallowed opinions, such as that nature was created in the image of God (as defined by man) or created from the sole perspective of what is best for humans (men and women).

Flat earth? Sun goes around the Earth? Human race and all the universe came into creation exactly in the year 4004 B.C.? Many angels will fit on the head of a pin? Man is at the center of the universe? No rings around Saturn? Forget it. No more. Those narrow views have mostly disappeared.

Even a simple scientific point that seems self-evident might lead to difficulty. E.g., how do we choose a "standard" day? Do we take a self-centered, human, and obviously culture-bound view? Or do we take a broader perspective?

(1) Specifically, do we take the anthropocentric position and say our conventional day should be based on solar time? One day = 24 hours.

(2) Or should a conventional day be defined in reference to the more universal sidereal (star) time? One sidereal day = 23 hours 56 minutes.

Chapter Six
Byways and Offshoots

Is time money?

A set of curious questions arises about time and motivation. What is the purpose of time? Why do we even want or need time? Why do we pay attention to it? Why define it?

A related question is this: If our rigid time line is so artificial and contrived, how did it grab the public's attention?

Who made us so smitten with commercial, industrial, standardized time? When did we first buy and sell time? Why do we persist in honoring the clock? Why do we seem to like objective time? Or are we simply dependent on it? Do science and technology now require clock time? What are the vested interests? Let's briefly chronicle how time became a ***commodity.***

The Benedictine order was founded in the sixth century. (Rifkin, p. 81) These monks didn't waste time. They kept busy. They made efficiency (using time wisely) into an art form as well as a moral and ethical imperative. They invented schedules to follow — to guide their pious activities.

> To ensure regularity and group cohesiveness the Benedictines reintroduced the Roman idea of the hour. (Rifkin, p. 84)

Mechanical crafts and expertise developed in Europe over many centuries in the Middle Ages. Clockmakers eventually rose to new heights. They created huge timepieces that captured imaginations. In a medieval town, consciousness of the clock's soaring architecture partially replaced that of the gothic-spired church. A new timing structure reached toward God. A towering town clock presided like a watchful beacon over daily life on cobbled streets.

Gradually, the brightest minds in Europe developed a world-philosophy that said the nature of our universe is like a clock. God created and set into motion a grand, clock-work scheme of nature.

> The most famous of the early clocks was one constructed at Strasbourg... Begun in 1547, it took twenty-seven

years to complete...it was over sixty feet high. (Levine, p. 90)

In a slightly more modern era Ben Franklin once advised, "Remember that time is money."

In the middle 1800s, the Industrial Revolution began to fill an objective gap in time. Time then was haphazard and localized.

> [The new machine age produced]...a few entrepreneurs who recognized the potential for marketing "time" as a product. (Levine, p. 65)

One of the American entrepreneurs [Langley]:

> referred to local time as a "fiction"... and went out of his way to proselytize for standardization. (Levine, p. 66)

Another of the new time-pushers [Waldo]:

> ...took the moral high ground...in a report to railroad commissioners about the needs of factory workers. "Any service which will train these persons into habits of accuracy and punctuality, which will affect all... with the same strict impartiality, so

> far as wages for time employed is
> concerned, will be a great benefit..."
> (Levine, p. 66)

Taking Waldo's lead,

> one of the primary marketing strat-
> egies of the clock companies was to
> promote the integrity—the inher-
> ent superiority—of clock time itself.
> (Levine, p. 67)

Personal time-pieces and clocks slowly became more affordable. Pocket watch popularity continued to climb. The time device business became, in a few short decades, *a big-money business.*

Finally, companies went so far as to argue that clock-timeliness or punctuality was a *moral and social* issue.

> The International Time Recording
> Company (later to become IBM)
> published a catalog in 1914. It said
> that time clocks "would save money,
> enforce discipline and add to the
> productive time...There is nothing
> so fatal to the discipline of the plant,
> nor so disastrous to its smooth and
> profitable working as to have a body

of men irregular in appearance, who come late and go out at odd times…"

> [The new time recorders] promised to help **"weed out these undesirables."** (author's emphasis) (Levine, pp. 68–69)

With this kind of judgmental attitude, capitalist investors, employers and labor bosses put time directly into the rough-and-tumble of politics. Resistance to this new time attitude made any rebel or nay-sayer subject to moral condemnation.

Is there a politics of time?

Some persons might innocently ask if time has become a problem, a controversy, or perhaps a new opportunity? Yes: it has indeed become all of these. We would do well to pay attention to recent books and opinions like those in Rifkin's 1987 *Time Wars*. Rifkin said:

> As society at large careens toward the high-speed culture of the twenty-first century, small pockets of protest have begun to appear at scattered outposts along the way to wage battle against the accelerated time frame of the modern age. (Rifkin, p. 4)

Those protesters

> ...would ask us to give up our preoccupation with accelerating time and begin the process of reintegrating ourselves back into the periodicities that make up the many physiological time worlds of the earth organism. (Rifkin, p. 4)

He goes on to talk about a new dynamic. He argues that this war is becoming a *huge,* ideological struggle. He sees a

> ...shift in the political spectrum away from the traditional spatial metaphors of right and left wing to a new temporal spectrum with empathetic rhythms on one pole and power rhythms on the other. (Rifkin, p. 7)

Can we compare apples and oranges?

It now makes sense to ask the following questions: To what would "objective time" apply? To what, or when, would "psychological time" apply? How should we approach these questions?

Mathematical rules can lay out just this kind of "field of applicability."

For example, we intuitively think that [2 apples plus 3 oranges] are not the same thing as [1 papaya and 4 mangos]. But, we also know that both groups do fit the category of [5 fruits]. *In the first sense or meaning the two groups or sets of objects are not equal. In the second sense, they are equal.*

The answer depends on how we pose or understand the question.

Hence we can correctly say, depending on the meaning or the circumstances, that this algebraic equation *[2A + 3O] = [1P + 4M] is both true and not true.*

Science often claims to be objective, regardless of our biases, human error, inadequate language or other problems with understanding. Science claims to be valid and true, *but within its own sphere.* Science usually acknowledges its limitations and what is outside its own sphere. It can recognize ways it does not apply, and places where it cannot go. Three examples of the latter are "belief in God," ritual observance and "superstition."

Here's a hunch—a guess—about how "time" began

Can we digress a moment? Go back a few millions of years. Try to imagine things as our very distant ancestor primates saw them. The earliest humans undoubtedly noticed that one day followed another. Will the sun rise tomorrow? The answer for them was Yes. They could assume the sun would rise (appear) again. Daylight follows dark. Then we go back to night, again and again.

Also, it would have been hard for those earliest humans to not notice change.

Change was surely coupled with growth, especially the growth of a new human being into a youth and then to an adult—to full size. Finally, that growth and change process would *also* lead to the distinguishing characteristics of old-age, decay, disability and death in some of the more senior citizens. Humans surely were aware of their mortality from the natural cause of age—from something mysterious about the aging and changing of life, in and of itself.

A conclusion or deduction to draw from these facts, in regard to the personal fates of those humans, was probably: "Our days are numbered."

> This sense of insecurity [from awareness of time] is a basic motivating force in religion and finds various forms of expression. (Brandon, p. 617)

Another conclusion was probably: Age or duration (our modem word) can be important. Duration was most likely measured in a primitive way by some of those earliest progenitors.

To summarize, the following assumptions might have been drawn by the early people: They could see both

- a sure thing/constancy (reliability) and

- importance (usefulness) to a generalized time sense.

This awareness seems very touching, human and cultural. This is doubly true when the time sense handles specific concepts like Father Time or the Grim Reaper (or any figure, symbol or thought related to death due to old age). But let's not forget that elephants also seem to grieve the deaths of family members. A developing prehistoric sense of time would not necessarily be strictly a **human** consciousness.

But this is all vague and fuzzy. None of these likely developments would support the appreciation of time as anything ***other than an abstraction***. Analysis of early humans certainly could not prove time was (is) an actual entity, nor a substance, nor even something "real" in the sense that food and hunger are real. The nature of time remains elusive.

Can a simple gimmick define time?

Is there a standard unit of time? How about a year? Or a day? After all is said and done, it may come down to a single day. Perhaps Day defines time's grand purpose for us. It may be that simple.

The easiest way to think of Day is as a "period of light." (Ferguson) Now, can you agree upon an average, standard or normal day? That has challenged a lot of scientists. OK, let's say you do agree. Then, all you have to do is chop up that day into little pieces. Presto! Abracadabra! Voila! **You have made time.**

You atomize, slice and reduce a fictional, arbitrary whole duration (the Day) into little pieces of duration. Say all pieces are equal. Say each shares the same quantity and quality. Keep track of them. Number them. Count them. Add and subtract them. Subdivide them. Now you can call this exercise the "measuring of time" or "math of time" or "keeping of time."

Such arbitrary, reductionist process yields the same kind of result when measuring linear distance: length. To begin, let's use a typical human "foot" as a length standard, or a meter (metric system), or whatever else is handy. Now, take it from there. I think of this process as a kind of mitosis of the mind. It's a divvying up something.

Is there an arrow of time?

We have been conditioned to react with amusement to the idea that time might go backward. Of course we say "Time marches onward." Time only goes one way. It's so very obvious. It's just common sense. So we believe.

Yet, science fiction stories have popularized the ideas of time travel, time going backward, and so

on. Some persons believe in circular and repetitive, pervasive and omnipresent, or holistic time.

Cosmology

One could argue that *the nature of motion* determines time. If it does, we have to look at the whole of the universe. Look at big-picture, cosmology theories. We'd have to ask a couple of elementary questions: "Why does the cosmos move?" And "How does matter in the universe move?" And a related question pops up here: "Can we predict motion in the future?"

The answers to these basic questions must be correct and true and give us knowledge, if we are to understand time. If our explanation or theory of the cosmos is wrong, then we'll also miss the point of what time is. It seems we are treated quite often to a new "Theory of the Month" as to what the universe is up to. Fact is: no one knows.

Chapter Seven
End of the Story

Let's now close by asking again: What is time?
It should be obvious to the reader that time is surely not what it sometimes appears at first glance. Perhaps, as the old saying goes: You can't judge a book by looking at its cover.

I have tried to explore one aspect of how we understand time, which is that time is an *art.*

Time seems to pose a complex philosophical question without a direct answer

Who might be able to answer? Can science tell us exactly what time is? Not yet. But we mustn't stop trying. We should continue to ask questions. Experiment. Research. Discuss.

Can religion tell us what time is? I doubt it. Not now anyway.

Can anyone? I don't think so. Something very deep and profound seems to be at work here. We should submit humbly to our present ignorance. Stand in awe. Then, continue to explore with an open mind.

It *could* be the case that...

> ...time is more than just a feature of reality; it may well be underlying reality. (Rifkin, p. 34)

Can we break the grip of clocks?

Intellectually, I believe that in today's world we are mostly ready and receptive to new ideas about time. This could lead to breakthroughs in science and philosophy. By releasing the hold absolute time has on us, we could clarify some modern dilemmas in physics and astronomy, make advances in other fields of study and knowledge, improve the quality of the lives of all people on our planet,, better understand other societies, shed light on God and the life of the spirit, enhance popular culture, and simply have fun.

Let time slip a bit. Think outside time's box. Stray off the straight line. Encourage new ways of speaking about time — of "tick talk." Give the science of time a little break. Take a breather. That is my appeal.

In conclusion

Beginning with three radical thoughts, I have tried to shake the rattle of how we look at time. I have attempted to clarify two kinds of time: objective and psychological. I have described how our cultures devise their own specific ways to live with, and to keep track of time. I have tried to synthesize our many views — our understanding — of time. I have tried to paint and tint the art of time.

By using this more complete and comprehensive picture, we could now move toward a new theory of time. Or perhaps not. Are we ready for that?

At least, as Levine has argued:

> Our goal should be to try to live in a "multi temporal" society, one in which we learn to move back and forth among nature time, event time, and clock time. In other words, each of us must chart our own geography of time. If we can do that, we will have achieved temporal prosperity. (Levine, back cover flap, publisher's note)

Appendix 1

(Quoted definitions are from Merriam-Webster.)

Definition 1:
Time may be *a vessel; a container*. Time just *is*. It's *a priori*, a given environment, as is water to a fish. Standard definition: it is "a period during which an action, process or condition exists or continues."

These things (action, process, condition) are here defined by their relation to the vessel of time.

Analogies are: a book that contains written words; a riverbed and riverbanks that contain flowing water; air that retains heat.

Does this imply 4D "space-time" depending upon mass, matter, energy and motion? Or are time and space more like 1D and 3D containers, like empty boxes, within which "things" exist and happen? Are the 3Ds or 4Ds of space/time like backdrops for a drama? I.e., are they a stage or structure within which, through which and upon which the story of existence unfolds and players act? Or with a more esoteric slant: Is our material world actually a wispy mirage, manifestation or mirror of some "more real" spiritual realm of existence? Deep philosophical questions swirl around this definition of time.

Definition 2:

We refer to a specific point in *time or moment* something happens, as in the case: "Your favorite TV show is on tomorrow at 6 PM." Or we may use the word "now." We capture the moment in the phrase, "arrived on time." Or time is — an identified moment, hour, day or year as indicated by a clock or calendar, as in "what time is it?" or "what day is it?" Or time is one of a series of recurring instances (moments), as in "how many times."

Definition 3:

Time is *an occasion,* as in "the last time we met." Also defined as "a point or period when something happens." This would include our term "time-sharing." An example: simultaneous access to a computer by many users.

Definition 4:

A *period of time,* especially with respect to some action or project is a "time frame." The word "during" means "throughout" or "at some point in." Example: "someone broke in during the night." A time period refers to the separation between, or distance between two points of time.

Definition 5:
Time is *a person's experience* during a particular period: "had a good time at the beach."

Definition 6:
Time is *leisure,* as in "found time to read."

Definition 7:
Time is *an age, or state of affairs.* Time is an era: The Roaring '20s are remembered as a rambunctious time. Time is a state of affairs or conditions: as in "hard times."

Definition 8:
We use *"when" or "then" words* to refer either to a point in time, or to distance between two points of time.

Definition 9:
We refer to *position,* either as *place or time* words — "after" and "before." Examples: "the third jogger runs before — ahead of — the fourth jogger." (place) And, "Saturday comes after Friday." (time)

Definition 10:
Some things endure — and some do not. We notice a contrast between things that last and things that do

not last. Some things happen slowly; others happen quickly. Examples: a favorite toy, made of durable material, delights us for many years.

And likewise, a gorgeous sunset fades away quickly after a few minutes. Also consider the phrase, "time-honored." A thing long-lasting can be "time-worn." We refer to timeless or perhaps eternal; as in "a timeless work of art."

Definition 11:

We **suspend** time. In an athletic game we have "time-outs." We use the expression "time warp" to mean "an anomaly, discontinuity, or suspension held to occur in the progress of time." The words "tarry" or "delay" can mean "to stop, detain or hinder for a time." The word "interrupt" is the opposite of "continue," which means "to remain in a place or condition."

Definition 12:

Many events in life require **timing**. If I am to meet a friend, we must arrive at the same place at the same time. If that agreement is planned and pre-arranged, then timing is of the essence. One could argue another point of view about encounters: in some cultural situations, such meeting of friends will just "happen naturally," be unplanned, be casual or will happen "if it's meant to happen." Another example of **timing** is play on a basketball court. You have to coordinate your ball pass ahead of a running teammate in such a way that the teammate catches the ball. Similar coordination requires pacing and coordination skill in music performance. We're very familiar with the idea of **bad timing,** such as the car driver who unsafely runs a yellow-turning-red light in traffic—and perhaps crashes. A related idea is **to schedule** something, as in "times his phone calls conveniently." Or, one uses a bus or train "time-table." The phrase "time bomb" refers to a timed and planned, or unplanned event.

Definition 13:

Examples of **repetition** from our daily lives enforce **a time sense.** Changes may be cyclical or repeating. Patterns, habits **and** expectations form. In the non-tropics, we notice green things sprout and grow in the springtime and die off in fall and winter. We notice the same calendar dates year by year such as February 14, July 4 or January 1. We go to sleep and wake in the morning as a habit, "like clockwork."

Definition 14:

Time is **a system** of reckoning time, as in "Mayan calendar" or "solar calendar."

Definition 15:

Time has passed. We may get a very strong, emotional sense and reminder of elapsed time when a death happens, or a divorce, or birth, or when we see someone we know who has aged markedly in appearance (looking "older") since we last saw them. We often hear words about a child: "My! How you've grown!"

Definition 16:
We ***add quantities***, as in "five times greater."

Appendix 2

Dimensions are arbitrary.

Dimensions don't really exist at all. They are constructs, imagined. We must accept them as ***given***, not as real. You can't separate space from time. Nor can you subdivide space arbitrarily into exactly three "dimensions." That subdivision has no connection to the real world. It's a mathematical convenience only. Nor can you say any one of the so-called "four" dimensions exists in an actual sense, because one dimension cannot be isolated from the others. We cannot consider one or more of the Ds separately, independently, without looking at the entire continuum.

Note that the only well-known analog found in nature for the 90-degree angle, or 3D model of structure, is the boxy shape of NaCl (salt) crystals. Caveat: Even though 3D and 4D are ***arbitrary,*** it is important to affirm that they are very convenient and ***useful.*** 3D and 4D dimensions can help us "frame" or "picture" the universe and what exists. Making straight, level lines and right angles certainly helps us frame a rectangular house. And those lines and angles make it possible to sketch a blueprint picture.

Examples of arbitrariness

Example One: Let's say we want to argue by example how both 3D and 4D are arbitrary. We could do the following. Mathematically construct a kind of "two-dimensional" space to explain space and volume.

This 2D would not be what we normally think of: i.e., it's not "width x length = area." Rather, this 2D (first example) would create *a field* or *sphere*.

1. Draw: a point. Think of that point as the *first,* or one dimension.

2. Then draw (or construct): other points at equal distances from that point. These other points describe the *second* dimension. The second point is differentiated from the first point, and that gives it a second dimension.

3. If drawn on a piece of ordinary paper, this exercise would draw a circle. But given free motion to draw (construct) in any direction, this 2D geometry would map and describe a sphere as the sum of all points equidistant from the first point.

Example Two: For yet another example of a different 2D space, create that space in the following manner.

1. Draw: a point. That is the *first* dimension.

2. Now draw: another point at a distance away from the first point or removed from the first point in time.

3. Now move something solid (having mass) from the first point to the other. Motion (between the two points) could be thought of as the *second* dimension. The speed, force and path of this "second dimension motion" would not figure into the definition.

4. Measurement: The "second dimension" would be simply the movement of something from point A to point B (assign a number value greater than or less than zero to represent this motion), or an object at rest between time A and time B (absence of motion, or zero value).

A final thought

Time is a circus, always packing up and moving away. (Hecht)

Bibliography

Barnstone, Willis, ed. *The Other Bible.* Harper & Row, 1984.

Berlinski, David. *Newton's Gift.* Simon and Schuster, 2000.

The Book. Tyndale House Publishers, Wheaton, Illinois, 1984.

Brandon, S.G.F., ed., *Dictionary of Comparative Religion.* Charles Scribner's Sons, NY, 1970.

Carroll, Sean M. "No Year's Eve?" *Smithsonian* magazine, January 2015, pp. 12–14.

Ferguson, Gerald R. *The Daily Crossword.* In the Jan. 15, 2003, *Oregonian* newspaper, Portland, OR.

Fraknoi, A, Morrison, D and Wolff, S. *Voyages Through the Universe.* Saunders College Publishing (a division of Harcourt), 2000.

Harrison, M. and Stuart-Clark, C. *Time Poems (the Oxford Treasury of).* Oxford University Press, 1999.

Hetzner, C.N. *In the War for Peace.* Moyer Bell, 1997.

Holling, Harriet Gilbert, spoken to the author January 2015, in Portland, Oregon

Hecht, Ben. http://www.brainyquote.com/quotes/quotes/b/
benhecht163774.html

Inglis, Stuart. *Planets, Stars and Galaxies.* John Wiley & Sons,
N.Y., 1967.

Kennedy, X.J. Quoted on National Public Radio, December
6, 2002.

Kent, Roman, "Updates on Holocaust Memorial Day as
events take place across Europe to commemorate 70
years since the liberation of Auschwitz from the Nazis,"
The Telegraph, Wednesday 28 January 2015; http://
www.telegraph.co.uk/history/world-war-two/11371241/
Holocaust-Memorial-Day-commemorations-across-Eu-
rope-mark-70th-anniversary-of-Auschwitz-liberation-lat-
est.html

Levine, R. *The Geography of Time.* BasicBooks, Harper Collins
Publishers, 1997.

Mailer, Norman. *Oswald's Tale.* Random House, 1995.

McCullough, David. *John Adams.* Touchstone / Simon &
Schuster, 2002.

Merriam-Webster Dictionary (paperback), 2001.

Morgan, Marlo. *Mutant Message Down Under.* HarperCollins,
1994.

The New Encyclopedia Britannica, Macropaedia, Vol. 19,
1988.

Prideaux, Humphrey. The Old and New Testament
Connected in the History of the Jews and Neighbour-
ing Nations. For Daniel. Earl of Nottingham, London,
Volume II, Part I, 1718.

Ridgeway, Rick. *Below Another Sky.* Henry Holt and Company, 2000.

Rifkin, Jeremy. *Time Wars.* Henry Holt & Company, 1987.

Setek, W.M. and Gallo, M.A. *Fundamentals of Mathematics.* Prentice Hall, 1999.

VanderKam, James C. *The Dead Sea Scrolls Today.* William B. Eerdmans Publishing Company, 1994